# BISBEE
## ARIZONA

# Yesterday & Today

### A Comparative View of the Queen of the Copper Camps

Written by Neil Bush
Photographed by Boyd Nicholl
Designed by Ruth Bush

AF684370

This book copyright © 1992 by Bisbee Image
All rights of reproduction, total or partial, by any means, are strictly reserved.

Library of Congress Catalog Card Number 92-90267
ISBN 0-9633014-0-3

Published by:
Bisbee Image
7 Howell Avenue
P.O. Box 1145
Bisbee, Arizona 85603
(602) 432-3332
Printed in U.S.A.

## CREDITS AND NOTES ON THE PHOTOGRAPHS

All of the Today photos were taken by Boyd Nicholl, a resident of Bisbee, whose enthusiasm for the project was unbounded. It was a pleasure working with him. The Bisbee Mining and Historical Museum (BMHM) and the Arizona Historical Society (AHS) provided all the Yesterday photos. Tom Vaughan, curator of the Bisbee Mining and Historical Museum, provided additional background on the Yesterday photos. Other sources of information were: "Bisbee: Queen of the Copper Camps," by Lynn Bailey, Westernlore Press, Tucson, Arizona, and the Bisbee Historic Walking Tour brochure, a cooperative project of the BMHM and the Bisbee Chamber of Commerce. My wife, Ruth, was responsible for the design and layout of the book and handled the production. Many thanks to all.

## Foreword

The idea for this book came from seeing similar publications in other areas of the country. Bisbee provided the perfect setting for a comparative view. The town is unique in that many of the buildings in the original photos are still standing and nearly all the photos could be duplicated by standing at the exact spot today! Fortunately Bisbee is not a developer's paradise.

For those of us who live here and love Old Bisbee the challenge of historic preservation is always before us. It is critical that we maintain the historic charm of the place and preserve the historic structures. A major problem facing us is the current deterioration of many buildings. Speculators who care nothing for the town own many of these properties and refuse to renovate, rent, or sell them. If this trend is allowed to continue Bisbee will not prosper and our town will slowly decay around us. Many positive things have been taking place and I hope this book will encourage those who agree that we should have better control over the direction our city is taking, to seek long term solutions to our problems, rather than the quick fix. I would have liked to point out many other details in the short captions, but if a picture is worth 1,000 words, these pictures should speak clearly enough.

Neil Bush April 1992

Doing this project with Neil and Ruth has been both enjoyable and surprisingly challenging. Photographically, it is quite difficult to duplicate another's vision. I had to not only find from where the original photographs were made, which was not always possible because of changes in our town, but when, in terms of the light, and what focal length lens to use to duplicate the original shot.

Of the times, what strikes me most forcefully about the difference between then and now is how few people are on Bisbee's streets today as compared to then. From my photographs Bisbee almost seems a ghost town, which we who live here know is not true at all! Our era simply does not live outside anymore. For even on a busy day, I never saw the crowds of people on the streets that are in the early photographs.

I also noticed how much of Bisbee we have already lost. Look at the details in the old buildings, gargoyles, porches and columns which are gone in the present. We have removed much of what was beautiful and replaced it with parking lots and huge highway signs with very little concern for beauty or history. And maybe more importantly we have forgotten our need for community which you can see in the randomness of the modern changes.

On the positive side of changes, look at the hills, covered once again with oak, juniper and manzanita. Without the old smelter and wood heating we no longer need to strip the hillsides for fuel.

Interestingly all the photographs were made with fairly extreme wide angle lens. I know the early Bisbee photographers were using view cameras for the most part, but I didn't realize they were using wide field lenses. Most of mine were made using a 28 mm lens on 35 mm film or a 90 mm lens on 4″ x 5″ film. You will notice that while the view is usually quite the same in both photographs then and now; due to the different focal length lens used, the relationship of the forground to the background is not the same. To absolutely duplicate an old photograph I would need to use the exact same lens and film size as well as be in the correct place.

One will notice the difference in tone values between then and now. Early films had a completely different response to the colors of the scene than what we have now. I did try, by the use of blue and green filters to bring the two closer together, but not all together successfully. The films like the times are just too different.

Boyd Nicholl April 1992

**Castle Rock 1888,** *BMHM Douglas Collection*
A journey through Old Bisbee must begin with Castle Rock, one of Bisbee's most easily identifiable landmarks, located on Tombstone Canyon. The original source of water in the mining camp and location of Bisbee's first settlements, Castle Rock could probably be called Bisbee's birthplace. In this photo the two small cabins below the rock were the residences of two early prospectors.

**Castle Rock c. 1906,** *BMHM Massara Collection*
It wasn't long before many miners' shacks sprouted up around the base of Castle Rock. The rock and wood beam supporting wall for the two homes in the center right of the photo are typical of Old Bisbee. Many of these original walls and their remains still support homes in Bisbee.

Today a portion of the original rock and wood beam supporting wall is visible. It also appears that the small shack to the lower left may be original. The home to the upper left of Castle Rock beneath the water storage tank may be original.

**Castle Rock c. 1902,** *BMHM MacGregor Collection*
The structure to the far right of Castle Rock was built by Joseph Muirhead who came to Bisbee in 1900 seeking work as a miner. The building served as a higher class boarding house. During rain storms water would rush down the canyon creating dangerous flood conditions. Early residents appreciated the cleansing affect of these rains but feared the destructive elements of the flooding.

Today Castle Rock appears much the same. The flooding problem has been managed with a culvert ditch built beneath the road. The railings in the lower left identify the ditch. The Muirhead recently was a bed and breakfast inn and now is a private residence.

**Bisbee Cemetery c. 1900,** *AHS*
This was the site of Bisbee's first cemetery. Due to the heavy flooding in the early camp, the cemetery was removed from this location to Evergreen Cemetery in Lowell during the mid 1890s. Phelps Dodge Corporation built a park here for the city in 1917... reminiscent of a popular 1980s movie.

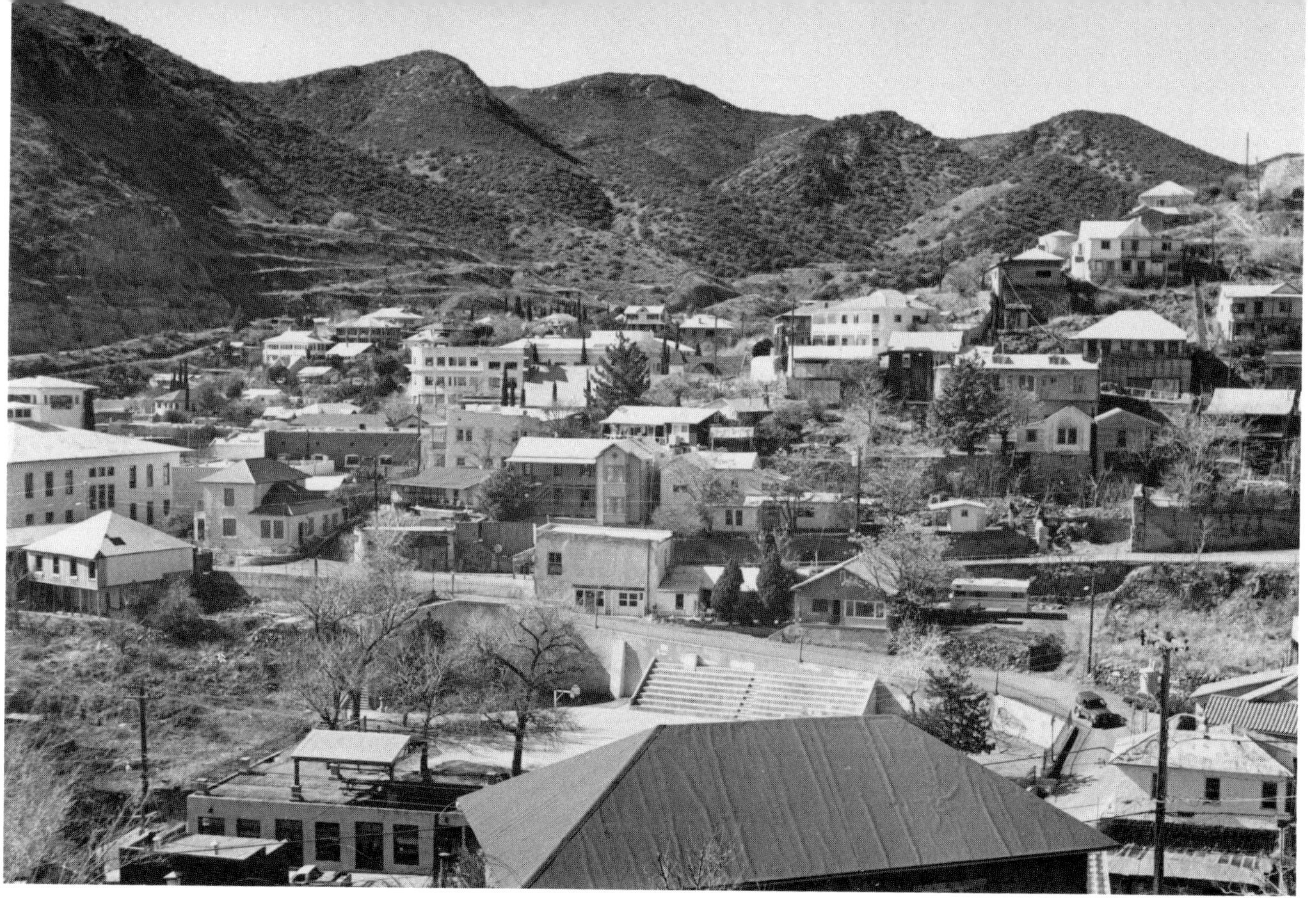

Today the original road which circled the cemetery still winds its way up and around City Park. Upon close examination it is possible to identify a number of original structures. The park is still a peaceful place as evident in this photo.

**Sacramento Hill 1886,** *BMHM Douglas Collection*
This early photo of Sacramento Hill shows a large and rapidly developing camp. Early smelter operations can be seen to the center left. All the early structures were of wood frame construction which would later be replaced by the sturdy brick structures which remain today.

Today Sacramento Hill is a mere memory. The small nub which remains identifies the edge of the Lavender Pit. Named after Phelps Dodge General Manager, Harry Lavender, stripping for the pit began in 1951. By 1955 the pit was ready for production and was responsible for extending mining operations in Bisbee until 1974. Will another pit be dug across the road? Rumors swirl.

**Bisbee c. 1904,** *BMHM Manley Collection*
This bird's eye view of Bisbee shows many of the classic structures. Easily identifiable are the Copper Queen Hotel (center), the General Office Building of the Copper Queen Consolidated Mining Company (left of hotel), the Copper Queen Mercantile (far left), and the edge of the Bank of Bisbee (right of mercantile).

Today the classic landmarks guide the eye to the town's later additions. The old Bisbee High School is located above the Copper Queen Hotel and Central School to the right. The general office is now the Bisbee Mining and Historical Museum and the mercantile has been replaced by the Bisbee Convention Center.

**Steps off Tombstone Canyon,** *BMHM Lockridge Collection*
In order to overcome Bisbee's mountainous terrain, sets of steps were constructed throughout town. Originally the steps were built out of wood as pictured on the left. Later, the steps were rebuilt out of concrete as detailed in the two other pictures. It's been said that a drunken miner was allowed three chances to climb the steps to his home before being taken to the city jail for the evening. In 1991, the Save Our Steps (S.O.S.) organization was formed to raise money for community

development. Every October the organization sponsors a unique foot race where runners race up the steps and back down the adjacent roads. The race is fast becoming a popular event and a significant fund raiser.

**Bisbee View c. 1903,** *BMHM State Capitol Museum*
This early photo shows many buildings which still stand. The Bank of Bisbee, the Presbyterian Church, the Y.M.C.A., the Copper Queen Hotel, and others. The current library/post office would later replace the structure in the lower left. The building on the far right was an early version of the Copper Queen Mercantile.

Today's photo details the numerous changes which have taken place over the years. Using the recognizable landmarks from both photos it is possible to find others which have survived.

**Bank of Bisbee c. 1902,** *BMHM Brophy Collection*
The Bank of Bisbee was established in 1900 by the mining establishment in response to Bisbee's rapid growth as a commercial and mining center. It was the first bank in Cochise County. The bank was probably constructed sometime between 1900 – 1902. The original organization would fall victim to the Depression. The building was acquired by the Bank of Douglas and later the Arizona Bank. Today it is Security Pacific Bank/Bank of America.

Comparing the two photos it is interesting to note the structural changes made to the building. The main entrance has been expanded, the lower windows widened, and an addition built. The bank was featured in the 1955 20th Century Fox motion picture, “Violent Saturday,” about a bank robbery in a small town. The movie starred Lee Marvin, Victor Mature, Ernest Borgnine, and Richard Egan.

**Bisbee View c. 1915,** *BMHM Crockett Collection*
This bird's eye view of early Bisbee shows a prosperous and still growing community.

Today much is the same in Old Bisbee. Take some time and compare.

**Brewery Gulch c. 1903,**
*BMHM Beckett Collection*
This striking photo shows a prosperous and bustling Brewery Gulch, "the livliest spot between El Paso and San Francisco." If Bisbee had a sign code in those days it isn't evident in this picture. The large stacks extending up the hillside provided draft for the copper smelter.

Today the Gulch is quite a different scene! It is interesting to note the facades on the right side of the street are still the same. Remains of the stone footings for the stacks can still be found on the hill side although not visible in the photo. The Gulch comes alive during July 4th when the traditional rock drilling and mucking contests are held.

**Muheim Brewery Building c. 1905,** *BMHM, Muheim Collection*
The Muheim Block was built in 1905 by Joseph Muheim, an early Bisbee entrepreneur. Joe began brewing beer in an old adobe structure which originally stood on this site. This activity was responsible for the street becoming known as Brewery Gulch. The building in the photo housed offices, the Edelweiss Restaurant, and later a stock exchange. Today the building is simply known as the "Stock" after the saloon upstairs which houses the original stock exchange board.

**Bisbee City Hall c. 1906–1907,** *AHS*
This building was Bisbee's first City Hall. Built at a cost of $13,000 in 1906 it was considered the finest city hall in the territory. The first floor was reserved for the fire department and contained stalls for horses. The second floor housed the City Attorney, Engineer, Clerk, Marshall, and Street Superintendent. City Council held their meetings upstairs. Until recent times the old city hall/fire station served as the Bisbee Police Station.

**Palace Livery c. 1910,** *AHS*
This large three story building was built after the fire of 1908. It housed the Palace Livery and still has an elevator which was used to hoist horses to the second floor stables. The window on the right appears to have the word "Undertaker" printed on it.

Today the building is the Johnson Gallery and is filled with paintings by local artists, pottery, Mexican masks, and many other interesting objects.

**Muheim House c. 1908,** *BMHM Muheim Collection*
The Muheim house was built by Joseph Muheim, an early Bisbee entrepreneur, sometime between 1898 and 1900. Located at 207 Youngblood Hill, the home is an example of late 19th Century architecture and has been restored to its former condition. Note the mule path running across the hillside in the background. It was probably used to transport fire wood and water.

The old trail is barely visible today. Little else has changed. The Muheim House is available for special events such as weddings, receptions, and other intimate gatherings. The house is open Friday through Monday 1–4 p.m. Contact the Bisbee Mining and Historical Museum for additional information.

**Tombstone Canyon 1906,** *BMHM Jarrett Collection*
This view, taken from atop Castle Rock, shows Tombstone Canyon during its early stages of development. It is easy to imagine the flooding which took place during summer monsoons by the condition of the dirt road. Amazingly, many of the homes in the photo are still standing today!

Today the road winds down around the canyon. Some of Bisbee's finest structures now appear including: St. Patrick's Catholic Church (center), the Cochise County Courthouse (left of church), and the County Administration Building (left of courthouse), which was formerly a school.

**Street Car Dedication 1906,** *BMHM Pickerell Collection*
In 1908, Bisbee was the largest city in the territory with a population near 25,000. This photo was taken during the Bisbee to Warren Street Car Dedication.

Today Bisbee's population is 7500–8500 people. The streets are not nearly as crowded, even during special events.

**Copper Queen Hotel c. 1903,** *BMHM Pavlovich Collection*
Founded to serve the interests of mining management, the Copper Queen Hotel was the most modern of its day. It took eighteen months to build at a total cost of $175,000. When the hotel opened its doors on February 22, 1902 it contained seventy-five rooms. There were forty-four guest rooms furnished with brass and iron bedsteads and large box spring mattresses. The kitchen contained all the latest appliances and the dining room seated seventy-eight persons. Other amenities

included a ladies' parlor, a billiard parlor, barber shop, washrooms, and bathrooms. Today the Copper Queen Hotel still serves guests from all around the world.

**Howell Avenue c. 1906,** *BMHM Crockett Collection*
This photo reveals that the Citizens Bank and Trust Company may have been located in the building in the right hand corner. The other awning indicates the entrance to a drug store. This entrance may have connected with the rest of the building providing banking, prescriptions, and other services at this location. The fence on the left is the same as today.

Today the Y.W.C.A. appears at the end of Howell Avenue, with large Italian Cypress trees growing along the street. A fire hydrant has been added to the steps.

**Parade of Volunteers 1912,** *BMHM Kentera Collection*
Brewery Gulch was the scene of many gatherings such as this parade of volunteers to fight the Balkan War in 1912. The Orpheum Theater (center) provided quality entertainment for many years.

Today there is far less activity in the Gulch. Unfortunately the theater no longer exists. A dentist still practices in the building to the right.

**Copper Queen Library/Post Office c. 1917,** *BMHM Moore Collection*
The fourth and final Copper Queen Library was built in 1906. The ground floor housed the post office which contained 2,500 boxes. There was no home delivery in Old Bisbee due to the mountainous topography, so the post office was a busy place.

The post office is still a busy place as is the library on the second floor. There is still no home delivery in Old Bisbee and the post office now contains 1,900 boxes.

**Central School 1906,** *BMHM Brophy Collection*
Central School was designed by architect F. C. Hurst and built in 1905. The design features a Romanesque portico capped with an Italianate bell tower.

Today two large cypress trees hug the bell tower obscuring the portico. The school houses the artist studios of the Central School Project.

**Main Street c. 1908,** *BMHM City Collection*
This photo of Main Street after a heavy rain and flooding shows a thriving commercial district. Uncle Sam's Loan Office probably served miners who were unable to secure conventional loans. At the end of the street is the Copper Queen Mercantile store destroyed by fire in 1938.

Today Main Street appears much the same structurally, especially the right side of the street. The Bisbee Convention Center, built in 1939, has replaced the mercantile at the end of the street.

**Loyalty Sunday Parade April 1917,** *BMHM Pickerell Collection*
This view of Tombstone Canyon was taken during the Loyalty Sunday Parade held in recognition of America's entrance into World War I in 1917. Note the Wrigley Chewing Gum advertisment in the lower right. The billboard is gone now but much remains the same. Even the water in the culvert ditch appears to flow in the same pattern as in 1917!

**Gymnasium Club (Y.M.C.A.) c. 1905,** *BMHM State Library Collection*
Considered one of the finest equipped Y.M.C.A.'s in the country, this building was contributed to the town by the active mining companies of the day. The idea was to combat the diversions of the liquor trade and stimulate production by providing healthy leisure time for the miners and their families. Today the building houses the Gym Club Apartments.

**Bisbee High School 1916,** *BMHM*

 The Bisbee High School was opened in 1914 and was partially destroyed by fire in 1919. It was reconstructed with the changes noticeable in these photos.

Today the high school appears much the same as it did after reconstruction in 1920. A parking lot has been added and the young tree has matured. The building now houses various offices.

**Y.W.C.A. c. 1916,** *BMHM Herbold Collection*
The Y.W.C.A. was erected in 1916, a gift of Phelps Dodge as a memorial to Miss Grace Dodge who at the time of her death was the national president of the Y.W.C.A. The building was the center of many social events.

Today the Y.W.C.A. provides rooms and board at reasonable rates and is used as a social center and for meetings by various local groups. A third floor was added some time ago.

**Bisbee Fire Dept. #2 c. 1914,** *BMHM City Collection*
Bisbee Fire Department #2, located up Tombstone Canyon, was built in 1914 and has been providing service ever since. Note the state of the art fire fighting equipment and the trolley tracks in the street.

Today the old fire department looks pretty much the same with good high tech equipment — well, compared to 1914 anyway!

**Drilling First Draft Sept. 6, 1917,** *BMHM Herbold Collection*
City Park was the center of many activities during its early years. The gentleman to the right was Harry Wheeler, Cochise County Sheriff, mustering inductees for the war effort in 1917.

Today City park is larger (note left of photo) but otherwise appears much the same. Basketball is today's most popular activity.

**Bisbee View c. 1935,** *BMHM Hoy Collection*
This is how Bisbee appeared about 1935.

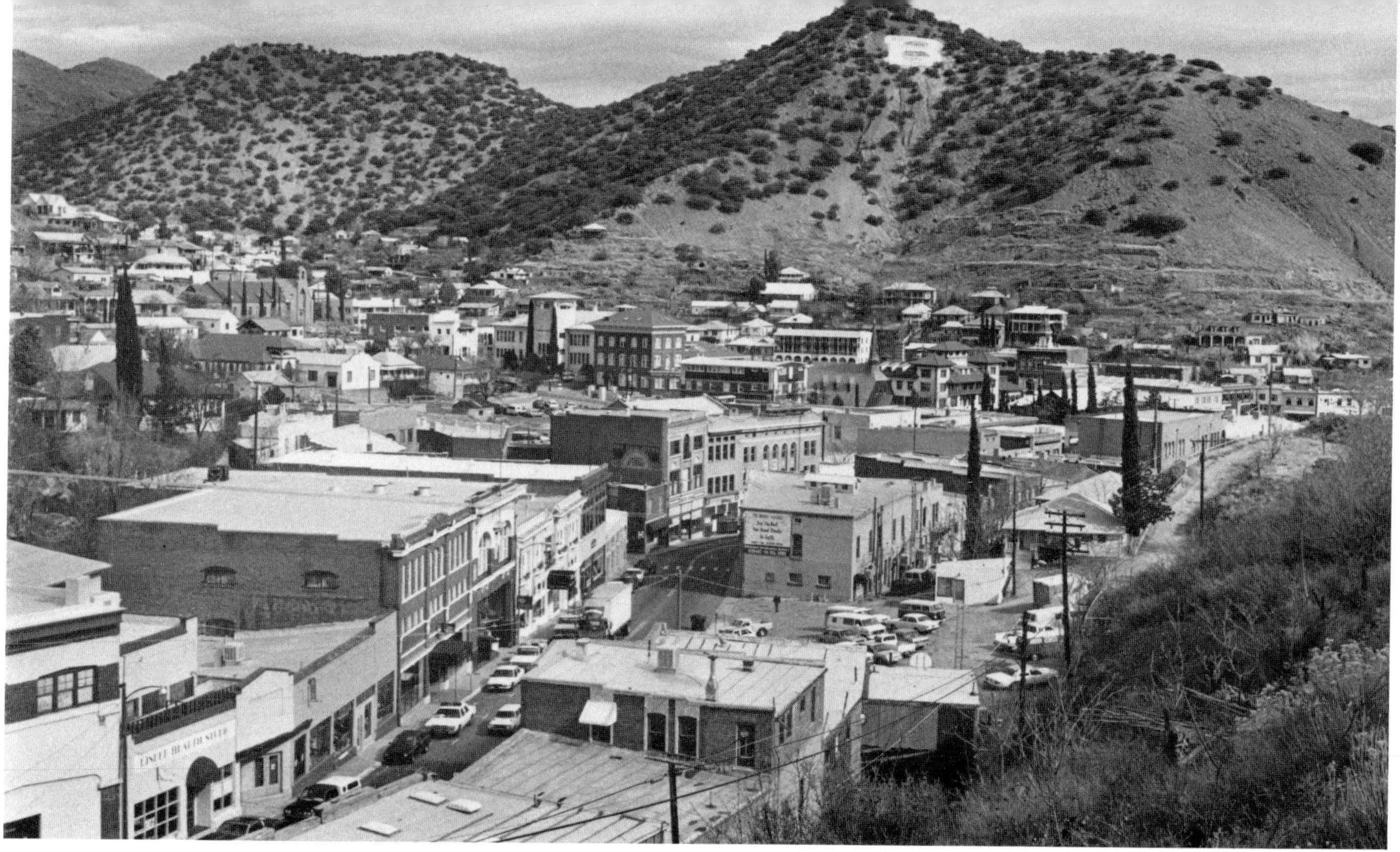

This is how Bisbee appears today.

**Phelps Dodge Mercantile c. 1940,** *BMHM Lee Smith Collection*
After the Phelps Dodge Mercantile burned to the ground in 1938 a young contractor named Del Webb was hired to build the replacement structure. This Art Deco project was his first. Later Del Webb became famous for his Las Vegas and Sun City projects.

Today the building is the Bisbee Convention Center. Community events are held here and the building houses shops and a restaurant/bar.

**Cochise County Courthouse c. 1937,** *AHS*
The Cochise County Courthouse was designed by Tucson architect Roy Place and built in 1931 at a total cost of $300,000. It is a wonderful representation of the Pueblo Deco style popular throughout the southwest at that time.

Today trees obscure the classic lines of the courthouse. The statue in the foreground is a tribute to the copper miners and was sculpted by local artist Phillip Sanderson. It was unveiled on November 11, 1935. Locals affectionately refer to the statue as the "Iron Man."